MEGAWATTS TO MEGA RECYCLING

Recovering end-of-life and landfill-bound
materials generated by large-scale
renewable energy projects.

BY

Emilie Oxel O'Leary

COPYRIGHT

**Megawatts to Mega Recycling© Copyright 2025
Emilie Oxel O'Leary**

ISBN

Paper Back.......978-1-964838-10-6
Hard Cover..........978-1-964838-12-0

DISCOVER MORE BY

EMILIE OXEL O'LEARY
www.emilieoleary.com

DEDICATION

To my loving and beautiful mother, Inga-Brita Oxel, who is now watching over me from heaven. You always believed that with God's strength, anything is possible. Your kindness to others continues to inspire me every day.

Because of you, I want to leave this earth a better place for my daughters — just as you did for me. I dedicate this book to you.

TABLE OF CONTENT

PREFACE

Renewable energy has transformed the world we knew into a power-rich beacon. Just a few decades ago, one could only dream about the notion of entire cities running on sunlight or wind. However today, that dream is our reality.

Solar farms stretch across long-distant deserts, and wind turbines rise from the seas, all to produce renewable energy. Together, these are the exceptional inventions of mankind solving our planet's most pressing crisis. However, it poses a visible obstacle in our race towards a cleaner and greener future. How? Due to the life cycle of such technologies, a pile of waste remains after a relatively short number of years, totaling a significant number of pounds of landfill waste.

As We Celebrate Solar Power, How Will We Manage The Eventual Waste From These Panels?

It's a question we can no longer afford to ignore. The renewable energy revolution has been driven by urgency, but it's time to pay attention to the mess that we are creating. Think about the future of these solar energy systems which have run out of their operational life. That's the turning point of Megawatts to Mega Recycling.

This book explores ways to keep renewable energy coming without causing too much waste. The book is not so much about solar energy's success as about the next step in that success:

The responsibility of managing waste materials. Take a step towards defining the true sustainability of renewable energy.

This book, in your hands, is both a road map and a call to take necessary measures before it's too late. Onboard the journey to learn all about the life cycle of renewable materials—from their inception during the solar boom of 2022 to their inevitable decommissioning, and the critical need to recover and recycle them. Each chapter dives into the challenges and opportunities of managing renewable waste, offering practical solutions for solar professionals, policymakers, and everyone invested in the future of clean energy.

As the renewable sector grows, so does the urgency to build a circular economy, one where the components of today's energy systems are repurposed for tomorrow. The chapters ahead outline not just the problem but a vision for how we can build a future where renewable energy doesn't just reduce carbon emissions but also minimizes waste.

The journey to a sustainable world is far from over. In fact, it's only just beginning. As we continue to expand our use of renewable energy, it's time to consider what happens next. I hope Megawatts to Mega Recycling inspires everyone to think about renewable energy not only as a means of generating power but also as a full-cycle process that honors both the planet and the people who rely on it.

The Mega Recycling Mandate:
Powering Beyond Waste.

FOREWORDS BY OTHERS

As a leading regulator who has approved the installation of millions of solar panels across Georgia, Emilie O'Leary is answering the questions I have been asking for a decade. What will we do with these panels when they have reached the end of their useful life? Her personal experience in running her company, Green Clean Solar, has given her the authority to "speak truth" to the renewable energy world and ask them to do better. It is not only the panels and materials in the field at the end of the project that need to be recycled, but the packaging material footprint too—as the panels arrive in cardboard boxes. Implementing the circular economy into the solar business should be intuitive, but it is not. Emilie explains the challenges, the costs, and objections, the inconvenience, the excuses—and of course the solutions to actually making solar recycling happen at scale. She is not dreaming because she does this for a living. No one has more authority in the solar recycling world than Emilie. I hope many of my regulatory colleagues read this book and create the necessary parameters to ensure solar panels don't wind up in landfills. Thanks, Emilie, for leading us into a better future.

Commissioner Tim Echols
Vice-Chair, GA PSC
Founder, Clean Energy Roadshow
www.cleanenergyroadshow.com
Host, Energy Matters Radio

TESTIMONIALS

Testimonial # 1

As a member of the US Green Building Council, I prioritize sustainable practices such as Recycling, Water Conservation, Waste Management, and Energy Efficiency. It's inspiring to connect with Emilie O'Leary, who shares my commitment to quality of life for both current and future generations.

Emilie's work ethic and dedication towards recycling education inspires companies to evaluate company's policies for recycling. She explains how it is possible to do the right thing for the planet with sustainable practices.

Both Emilie and I, as mothers, deeply consider the well-being of our children, grandchildren, and the resources needed for clean air, water, and land. It's heartening to find like-minded individuals who are dedicated to creating a more sustainable and healthier environment for our families and communities.

It is my opinion; women tend to think more about mankind's future more than men. Men tend to focus on the present, but women have the ability for the present while thinking about long-term future scenarios. More women should embrace a career in sustainability careers to provide framework on why we should care about the products shelf life.

Jennifer Quas
Account Manager Southeast
CPS America, Chint Power Systems

Testimonial # 2

Emilie Oxel O'Leary has been a champion and mentor for women in the Renewable Energy industry. She has been at the forefront of the industry, championing the implementation of recycling best practices aligned with the Renewable Energy mantra of sustainability and d-carbonization. Her experience as a female entrepreneur has enabled her to glean invaluable, first-hand insights to share with her industry colleagues. Emilie is a guiding light for a renewable, sustainable future that we all can embrace.

Solar Land Agent, LLC
Mary Ellen Barker, MBA
Founder & CEO

Testimonial # 3

Emilie O'Leary is tackling one of the solar industry's most urgent upcoming challenges—what happens to panels at the end of their life. As we scale solar globally, we must also lead with accountability and foresight. Emilie brings real-world execution experience and a no-nonsense approach to building the infrastructure and mindset for solar recycling. This book is a call to action for all of us to align clean energy with true sustainability.

Dan Shugar, P.E.
Founder and CEO
Nextracker

INTRODUCTION

In Megawatts to Mega Recycling, solar trailblazer Emilie Oxel O'Leary introduces readers to the intricate web of consequences that unfold with the surge in large-scale utility solar projects advancing d-carbonization. As an advocate for green business and a leading woman navigating the renewable energy sector, Emilie has observed firsthand how the expansion of solar installations, while heralding a cleaner energy era, simultaneously ushers in a growing tide of waste. She poses a critical question to readers: What will our future look like when mountains of solar panels end their service lives and are relegated to landfills? This book is a journey through the potential long-term effects and health concerns stemming from the solar waste dilemma, underscoring the imperative for effective waste management and recycling protocols to be addressed proactively. Reflecting on just one startling fact a total of 93,000+ cardboard boxes were generated from a single solar installation project managed by Emilie's company, Green Clean Solar. Emilie invites readers to examine the broader environmental implications of renewable energy's packaging material and waste footprint. This one standalone solar project statistic is a window into the larger issue of how renewable projects entail significant resource and waste challenges. [1]

Emilie invites us to compare a future where we ignore our solar waste problem, allowing landfills to be strewn with solar panels and other recyclables, versus a proactive solar industry that plans ahead and follows through with solutions to manage the aftermath of renewable tech waste.

The narrative of unintended consequences threads through the book as you examine the paradox where efforts to mitigate climate change through renewable energy could inadvertently fuel a waste crisis. This discussion includes an examination of the role of women in the renewable sector. With women receiving a paltry 7% of funding in the renewable energy industry, Emilie argues for the critical need to bolster female representation and leadership in environmental decision-making and innovation. Diverse perspectives are essential to crafting sustainable solutions to the waste challenges posed by large-scale solar projects.

Megawatts to Mega Recycling stresses the urgency of addressing these issues head-on, highlighting the future unintended consequences of the solar energy boom. It's a clarion call for proactive engagement with waste solutions today to prevent being caught unprepared as the volume of solar waste escalates. Megawatts to Mega Recycling is both an analysis and a proactive appeal, urging a comprehensive rethinking of how we plan, execute, and eventually retire renewable energy infrastructures.

Emilie sets the stage for the reader to reflect on the dual narrative of human progress and our ability to harness the sun's power, while contrasting our ever-evolving grasp of the environmental legacies we're creating. Through these words, she hopes to inspire a forward-thinking, circular economy approach in renewable energy practices, advocating for consideration of the entire life cycle of renewable energy technology implementation.

CHAPTER 1
THE SOLAR BOOM AND DECOMMISSIONING IMPLICATIONS

At the beginning of my solar career in 2016, landing a 1 MW project was a major accomplishment. The solar business was about to undergo a massive revolution, but its scope and rate of change were still not well understood. The next three years, from 2016 to 2018, marked an important point in the world's energy transition to renewable energy, as evidenced by the rapid expansion of solar installations in the United States.

A number of elements came together to generate the boom, each amplifying the other. Innovations in technology were critical. With the sharp decline in solar panel costs, solar projects became more financially feasible than ever before. Initial solar installation prices were drastically lowered by improvements in manufacturing techniques and the efficiency of photovoltaic cells. This cost drop was the result of a technological breakthrough that made solar energy feasible and accessible on a scale that was previously unthinkable, not just a matter of economics.

Government policies and incentives were equally pivotal in this period of expansion. Regions across the nation introduced ambitious renewable energy targets, backed by supportive policies such as feed-in tariffs, tax credits, and renewable energy mandates. These initiatives provided fertile ground for solar project development, catalyzing the sector's growth.

The surge in environmental consciousness and the pressing concerns over climate change also played a critical role. Businesses and utilities, driven by a mix of regulatory pressures and public demand, increasingly adopted renewable energy solutions. Solar energy, with its scalability and abundance, emerged as a particularly attractive option. High-profile companies, including giants like Amazon, FedEx, and Target, led the way, integrating solar energy into their operations and sustainability strategies.

Financial institutions, recognizing the long-term viability and profitability of renewable energy, began to channel substantial capital into solar projects, further accelerating their deployment. The industry's growth was not just about the energy produced; it was about a fundamental shift in how energy investments were perceived and executed.

During these key years, my mechanical installation solar business grew. What was previously hailed as a one-megawatt project gradually started to escalate. Megawatts grew into double and triple digits. Our portfolio now includes projects with a capacity of up to 100 MW, demonstrating the thriving market and the industry's rapid expansion.

However, as solar farms expanded, a new worry emerged decommissioning. End-of-life management of solar arrays became a significant issue. With typical lifespans of 25 to 30 years, the sheer size of the facilities demanded strict planning for eventual decommissioning and disposal. This period witnessed the development of rules and regulations for responsible decommissioning, such as recycling and disposal processes for solar panels and their components.

Despite these initiatives, there were few in-person talks regarding the end-of-life of large-scale solar projects. The industry's main priorities were building and moving on to the next project as quickly as feasible. The idea of sustainable life cycle management was still developing and was frequently eclipsed by the quick satisfaction of rapid deployment and construction.

Another watershed was reached in 2022 when the Inflation Reduction Act (IRA) was passed, stretching the limits of project sizes. The Act catalyzed $28 billion in investments toward domestic manufacturing of electronic components, solar panels, inverters, and batteries.

This was only the start of the legislative support. By 2035, renewable energy investment is expected to reach $1.2 trillion, according to industry analysts like Wood Mackenzie, indicating a period of unprecedented development and opportunity.

The manufacturing industry reacted quickly. Large-scale production facility investments started to appear all over the country, showing that the sector was prepared to meet the needs of the domestic market. These advancements aimed not only to increase capacity but also to create a key silicon-based solar supply chain in the United States. The scope of these announcements was indicative of the industry's commitment to fulfilling the dcarbonization goals put forth by the U.S. and the global community.

As the demand for solar projects surged, so did the production of solar panels. This led to an unprecedented scale-up of manufacturing facilities worldwide, designed to meet the escalating demand for modules. However, this growth spurt was not without its challenges. The increased production inevitably led to a massive generation of waste, particularly in the form of packaging materials. Wood pallets, cardboard, and plastic, essential for transporting and protecting solar panels, racking, and other equipment,

Contributed to a significant increase in the amount of waste generated. This unintended consequence highlighted a critical oversight in the industry's sustainable practices.

This realization hit me hard. As we built massive solar projects, we found ourselves complicit in generating considerable waste, often ending up in landfills. It was a stark contradiction to the clean, green image of the solar industry. This period served as a wake-up call, prompting me to reconsider the broader environmental implications of our work and how to lead the industry toward better practices. The industry's focus on rapid expansion sidelined the importance of sustainable waste management.

Exploring these waste impacts of solar energy installations revealed a multifaceted challenge, intertwining environmental concerns with economic opportunities. My journey through the solar industry has offered me a firsthand look at the repercussions and responsibilities that come with the sector's growth.

THE UNSEEN COST OF SOLAR EXPANSION:

When we analyze the entire life cycle of a solar project, from installation to decommissioning, I realized that we must acknowledge the unavoidable waste generated.

Rapid installation, while necessary for progress toward decarbonization, starts this cycle. The National Renewable Energy Laboratory (NREL) believes that solar panel packaging waste, which includes wood, cardboard, plastic, and scrap metal, accounts for a large amount of the waste.

As solar farms mature, the need for retrofits or midlife upgrades arises, resulting in additional waste and recycling opportunities. These upgrades, which attempt to improve efficiency or replace worn-out parts, produce a combination of electrical, metal, and packaging waste because new parts must be hauled in by pallet to replace old ones for maintenance.

The Solar Energy Industries Association (SEIA) emphasizes the critical need for effective recycling programs to manage this waste, underscoring the industry's responsibility in mitigating its environmental footprint.

Decommissioning marks the end of a solar installation's life, presenting perhaps, the most significant waste management challenge. According to SEIA, decommissioning utility-scale solar farms can result in the disposal of several hundred tons of materials, including metals, glass, and potentially hazardous substances like cadmium and lead from a small subset of solar panels. Without proper management, this waste can have detrimental effects on soil, water, and air quality, thus counteracting the environmental benefits of the solar energy it once produced.

ENVIRONMENTAL IMPLICATIONS:

Waste from solar installations has an environmental impact that extends beyond the projects' local neighborhood. If trash is not adequately managed, it can lead to landfill overuse, soil degradation, and water pollution, especially for large-scale installations. Waste incineration, a prevalent method of disposal, can emit harmful emissions, which contradicts solar power's clean energy goals. Even if all landfill efforts are successful in the short term, we cannot predict the long-term consequences of solar waste resting in a landfill. Could it affect human health in the future? Could it lead to large landfill space takeovers during decommissioning? These future considerations should be made today so that we can prevent unanticipated negative consequences.

Aside from the immediate impact, there's the issue of resource use. Solar panels contain important components such as silicon, silver, and aluminum, which, if not recycled, represent a major waste of resources. According to the lifespan analysis of solar panels conducted by institutes such as NREL, a complete recycling program may recycle more than 90% of these materials, gradually reducing the requirement for virgin resource extraction and lowering the environmental impact of solar production.

ECONOMIC AND COMMUNITY IMPACT:

The trash generated by solar arrays is both an environmental and economic issue. There is a growing opportunity in the waste management and recycling sector that is specifically geared to the solar industry. Creating a strong recycling infrastructure can boost job creation and economic growth, especially in areas with high solar development. Local communities can profit from the creation of recycling facilities, which can create jobs and boost the local economy in both the renewable and waste industries.

Furthermore, the solar waste management business might spur innovation in recycling technology, resulting in more efficient processes and even new markets for recycled materials. This can foster a circular economy where waste is viewed as a resource for new manufacturing processes instead of an end product.

SUSTAINABLE PRACTICES AND POLICY FRAMEWORKS:

The solar industry needs to incorporate waste management solutions at every stage of the project's lifespan if it is to grow sustainably. This approach begins with sustainable design, giving recycling top priority and reducing waste production. Proactive waste diversion techniques, like on-site recycling and material reuse, can be implemented throughout operations.

Policy frameworks play an important role in this scenario. Incentives that enforce recycling and proper waste management for solar systems can help ensure industry-wide compliance. Financial incentives for implementing such techniques can motivate solar firms to promote circularity.

Collaboration between the solar industry, waste management experts, and policymakers is critical to developing a comprehensive strategy that addresses the national waste challenges while supporting the sector's growth.

Collaboration between industry stakeholders, government agencies, and environmental groups is essential to developing effective policies and strategies for solar waste management. Initiatives like SEIA's PV recycling program demonstrate the industry's commitment to sustainable practices. Additionally, policy frameworks that encourage recycling and responsible waste management can further enhance the sustainability of the solar energy sector.

The waste generated at each stage of a solar farm's life cycle—from installation to decommissioning—poses significant management challenges. Addressing these challenges requires a concerted effort from the solar industry, policymakers, and environmental advocates to develop and implement effective waste management and recycling strategies. By prioritizing sustainability throughout the life cycle of solar projects, the industry can minimize its environmental impact and contribute to the broader goals of waste reduction and resource conservation.

A CLOSER LOOK AT THE STAGES
OF SOLAR RECYCLING:

The life cycle of utility-scale solar farms consists of several stages, each of which contributes to the total waste footprint of solar energy production. This topic focuses on the waste management challenges and environmental implications of solar farm installation, operation, and decommissioning.

INSTALLATION WASTE,
THE BEGINNING OF THE CYCLE:

The installation phase of utility-scale solar farms produces a substantial amount of waste, primarily from the packaging of solar panels and related components. These materials include a lot of cardboard, plastic wrapping, wood pallets used for shipping, and waste metal from construction projects. Given the size of these installations, which can span several acres, the volume of packing waste can be significant. The Environmental Protection Agency (EPA) emphasizes the significance of reducing packaging waste in order to save landfill space and resources.

Strategies such as returnable packaging, material optimization, and recycling can all help to reduce the environmental impact of this phase.

OPERATIONAL WASTE: MAINTENANCE AND UPKEEP:

During the operational life of a solar farm, regular maintenance, repairs, and occasional replacements of components generate waste. This includes not only the small parts like brackets and fasteners but also larger components like damaged solar panels, inverters, and support structures. For instance, a single damaged panel in a farm of 250,000 can seem insignificant, but on a larger scale, the waste generated from regular maintenance and repairs can accumulate significantly. The replacement of heavy components, such as shock absorbers and transformers, involves shipping new parts in packaging materials, further contributing to the waste stream with pallets, cardboard, and plastic.

DECOMMISSIONING WASTE AND END-OF-LIFE MANAGEMENT:

The decommissioning phase concludes the life cycle of a solar farm by dismantling and removing all equipment and structures. This method produces a significant quantity of garbage, such as solar panels, racking systems, cables, and concrete foundations. The problem is to responsibly manage this garbage, with a strong preference for recycling over landfill disposal. According to the International Renewable Energy Agency (IRENA), end-of-life management for solar panels is crucial to ensuring that the materials are recycled and reused, hence lowering the overall environmental impact of solar power.

The waste generated during each step of a solar farm's existence has both environmental and economic consequences. Waste that is not adequately managed can contaminate soil, pollute water, and increase greenhouse gas emissions. Effective waste management measures, including recycling and material recovery, can help to reduce these environmental impacts and contribute to a more sustainable solar sector. Economically, building a strong recycling infrastructure for solar waste can generate jobs, promote innovation, and reduce raw material demand.

WIND RE-POWERING AND DECOMMISSIONING CURRENTLY PEAKING:

The wind energy sector is going through a large period of re-powering and decommissioning, especially as many wind turbines reach the end of their operating lives. About 85% to 90% of a wind turbine's mass, which includes materials such as aluminum, steel, copper, and iron in the tower and nacelle components, is easily recyclable. The turbine's blades, which make up 6% to 14% of its bulk, are composed of composite materials such as fiberglass or carbon fiber, making recycling harder. Re-powering entails upgrading or replacing outdated wind turbines with modern, more efficient technologies. This procedure increases the productivity and sustainability of wind energy plants while also extending their lifespan. Partial re-powering, for instance, might improve certain parts, like gearboxes and blades, while keeping towers and foundations in place. This would boost the energy production and capacity factor.

Re-powering and decommissioning are prioritized because modernizing the current wind fleet is crucial to maximizing its contribution to the energy system and facilitating the change to more sustainable energy usage. The International Energy Agency Wind Technology Collaboration Program, spearheaded by NREL, has been actively exploring and supporting these initiatives, highlighting their benefits for the environment and the economy.

Efforts are underway to improve the recycling of wind turbines, with a focus on designing blades and other components that can be more easily recycled or repurposed. Initiatives like the partnership between GE Renewable Energy and Veolia to grind turbine blades for cement-making processes illustrate the industry's move towards full recycling. This transition to more sustainable practices reflects the wind sector's response to the challenges of waste management and environmental sustainability.

THE ROAD AHEAD:

As we advance in solar, the focus must go beyond maximizing energy output to include minimizing environmental impact, including waste. The industry must embrace a life cycle perspective, acknowledging that each phase of a solar panel's life, from production to upkeep and decommissioning, has environmental implications. Through innovation, collaboration, and regulatory support, we can address these challenges, turning potential waste into an opportunity for sustainable growth.

In my role, witnessing the solar industry's evolution has been enlightening and challenging. The waste impact associated with solar installations presents a complex puzzle, balancing the benefits of clean energy against the environmental and economic costs of waste.

However, with concerted efforts and sustainable practices, the industry can navigate these challenges, ensuring that solar energy remains a key player in the global transition to sustainable energy. The solar sector, in collaboration with regulatory organizations and environmental organizations, is aiming to promote more sustainable waste management techniques. Research and development activities are aimed at increasing the recycling of solar panels and lowering waste during manufacturing. For example, NREL's work on life-cycle assessment and PV recycling seeks to design solar products with end-of-life recycling in mind, supporting a circular economy approach to the solar sector.

STILL, THE ENDS JUSTIFY THE MEANS:

The momentum for change is obvious, with programs targeted at minimizing the environmental impact of solar panel manufacturing and end-of-life disposal. Solar is a cleaner and more sustainable energy source than typical coal-fired power plants, which generate over 110 million tons of coal ash per year in the United States, frequently resulting in hazardous leaks and spills. The solar industry's effort to tackling its recycling issue not only strengthens its position as a clean energy leader, but also aligns with global environmental aspirations. By continuing to innovate and implement responsible recycling management techniques, the solar industry can maintain its position as a cornerstone of the clean energy transition, demonstrating that renewable energy solutions can be both effective and environmentally friendly.

CHAPTER 2

CURRENT STATE OF WASTE MANAGEMENT AND RECYCLING

The waste management industry has long been a cornerstone of environmental sustainability, evolving over the decades to address the growing challenges of a world increasingly dependent on technology and consumer goods. From traditional landfill operations to advanced recycling techniques, the industry has continuously adapted to meet society's demand for responsible disposal. Today, with the rapid expansion of renewable energy systems, a new frontier has emerged: the recycling of solar panels.

Solar panels, while celebrated for their role in reducing carbon emissions, present a unique waste management challenge. Designed to last 25 to 30 years, the first large wave of installations is now approaching the end of its life cycle. As these panels reach obsolescence, the waste management industry is grappling with how to process and recycle the materials effectively, efficiently, and sustainably. This chapter explores who is stepping up to meet this challenge and how the industry is innovating to handle the growing solar waste crisis.

THE EVOLUTION OF SOLAR PANEL RECYCLING:

Several players in the waste management and e-waste industries are pivoting their focus to include solar panels. Historically, recycling facilities and scrap yards have concentrated on traditional consumer electronics, such as computers, mobile phones, and household appliances. However, as the demand for solar panel recycling grows, many of these facilities are adapting by investing in specialized equipment capable of processing solar panels. These investments reflect a recognition of the significant opportunity—and responsibility—to manage this new waste stream.

Companies specializing in electronic waste (e-waste) recycling are leading the charge. They are retrofitting existing facilities or building new ones equipped with advanced technology to recover valuable materials from solar panels. This includes high quality glass, aluminum frames, silicon, and even trace amounts of precious metals like silver and copper. With the right machinery, some facilities can now achieve material recovery rates as high as 95%, turning end-of-life solar panels into valuable commodities rather than landfill waste.

KEY PLAYERS IN SOLAR PANEL RECYCLING:

Across the globe, different organizations are stepping up to address solar recycling:

1. **Traditional Recycling Management Companies:**
   Many large recycling management firms are beginning to explore solar panel recycling. They are leveraging their logistical expertise and infrastructure to collect and transport panels to specialized facilities.

2. **Electronic Recyclers:**
   Established e-recycling recyclers are expanding their operations to include solar panels. By integrating new processes and equipment, they are adapting to the unique challenges of handling this type of recycling.

3. **Manufacturers:**
   Some solar panel manufacturers are taking proactive steps to manage the life cycle of their products. Through take-back programs and extended producer responsibility (EPR) initiatives, these companies are ensuring their panels are recycled in compliance with environmental regulations.

4. **Startups and Innovators:**
   Emerging companies are developing cutting-edge technologies to improve the efficiency and cost-effectiveness of solar panel recycling. These innovations include

chemical and thermal processes that enhance material recovery rates and reduce recycling.

THE TWO MAIN TYPES OF SOLAR PANELS AND THEIR RECYCLING PROCESSES:

The solar industry predominantly relies on two types of panels: crystalline silicon (C-Si) and thin-film panels. Each type has distinct characteristics and recycling requirements.

1. **Crystalline Silicon Panels:**
- **Composition:**
    o C-Si panels are made primarily of glass (about 76% of the panel's weight), aluminum (frame), silicon (cells), and smaller amounts of silver and copper.

- **Recycling Process:**
    o The aluminum frame and junction box are removed first for recycling.
    o The glass is separated and cleaned for reuse or recycling into new glass products.
    o Silicon cells are processed to recover silicon, which can be reused in new panels or other applications. Advanced thermal or chemical treatments are used to recover valuable trace metals like silver.

- **Challenges:**
    Recovering silicon and silver economically is a significant challenge due to the low market value of these materials relative to the recycling cost.

2. Thin-Film Panels:

- **Composition:**

Thin-film panels are made of a thin layer of photovoltaic material (such as cadmium telluride or copper indium gallium selenide) deposited on glass or metal.

- **Recycling Process:**
  o The panels are shredded, and the materials are separated using mechanical and chemical processes.
  o Metals like cadmium and tellurium are recovered and purified for reuse.
  o Glass is also recovered and recycled.

- **Challenges:**
  o Handling hazardous materials like cadmium requires specialized facilities and adherence to strict environmental regulations, adding complexity and cost to the recycling process.

CHALLENGES IN SOLAR PANEL RECYCLING:

While progress is being made, the industry still faces several hurdles:

- **Economic Viability:**
  The cost of recycling solar panels often exceeds the value of recovered materials, making it a less attractive business without subsidies or mandates.

- **Logistics:**
  Solar panels are bulky and fragile, complicating their collection and transportation to recycling facilities.

- **Regulatory Gaps:**
  Inconsistent regulations across regions create uncertainty and hinder the development of a cohesive recycling infrastructure.

PROMISING DEVELOPMENTS AND THE FUTURE:

Despite these challenges, the future of solar panel recycling looks promising. Governments are beginning to implement policies to support recycling efforts. For example, the European Union's Waste Electrical and Electronic Equipment (WEEE) directive requires manufacturers to take responsibility for the end-of-life management of their products. In the United States, states like Washington are enacting similar legislation to encourage recycling.

Additionally, advancements in technology are driving progress. Thermal, chemical, and mechanical processes are being refined to improve recovery rates and reduce costs. Some facilities are even experimenting with fully automated systems to streamline operations further.

The waste management industry is at a pivotal moment as it rises to meet the challenge of solar panel recycling. With e-waste recyclers, manufacturers, and innovators leading the way, the foundation is being laid for a sustainable solution to this growing problem. By fostering collaboration, advancing technology, and increasing public awareness, we can transform the issue of solar waste into an opportunity for environmental and economic growth—and pave the way for a truly circular economy in renewable energy.

IMPACT OF SOLAR WASTE ON ENVIRONMENT

During my research for Megawatts to Mega Recycling, I made it a point to visit various waste facilities across the country. One particularly memorable trip was to a popular landfill in California. With all the media buzz about solar panels being irresponsibly dumped into landfills, I wanted to see for myself what was really going on.

As I drove up to the facility, I couldn't help but feel a mix of curiosity and skepticism. The reports had painted a grim picture—discarded solar panels piled up, contributing to environmental degradation. If that were true, it would be deeply concerning given the renewable energy industry's commitment to sustainability.

Upon arriving, I approached some of the workers at the site. "Is it true that people are dumping solar panels here?" I asked directly.

Their response was immediate and firm: "We're not allowed to accept solar panels."

That answer was both surprising and reassuring. Given the stringent regulations around electronic waste in California, it made sense that solar panels would fall under strict disposal rules. Still, I was determined to see things with my own eyes. I drove further up the landfill, scanning the vast landscape for any sign of solar panel debris—broken frames, shattered glass, or even intact panels waiting to be buried.

But there was nothing remotely close to a solar panel in sight.

This visit was a pivotal moment for me. It contradicted the narrative I had seen in numerous headlines and articles. While media reports often highlighted the problem of solar waste ending up in landfills, at least in this instance, it was clear that the facility was adhering to proper disposal protocols.

Of course, this doesn't mean the problem doesn't exist elsewhere.

The solar industry still faces significant challenges in managing end-of-life panels. Recycling infrastructure is limited, and many facilities aren't equipped to handle the unique materials found in solar technology. However, my visit to this landfill underscored an important point: we need to be cautious about generalizations. Not every landfill is a dumping ground for solar waste, and some facilities are doing their part to follow regulations and protect the environment.

This experience also reinforced my belief in the importance of transparency and accountability within the renewable energy sector. If we are to maintain public trust and continue advancing clean energy solutions, we must ensure that our waste management practices align with the values we promote. Visits like this are crucial for separating fact from fiction and understanding the realities on the ground.

Moving forward, I remain committed to advocating for better waste management solutions and sharing the stories of facilities and companies that are getting it right. My visit to that California landfill was a reminder that while challenges remain, there are also successes worth acknowledging and building upon.

IMPACT OF SOLAR RECYCLING ON ENVIRONMENT:

Landfills are not kind to solar panels. These sites bring a plethora of problems that severely impact our environment. When discarded electronics and plastics break down, they produce toxins that can pollute groundwater and cause foul-smelling odors. Adding organic waste, like food, cardboard, and paper, to landfills generates a significant amount of methane, a greenhouse gas 25 times more powerful than CO2.

The impacts are long-lasting, as even after closure, landfills can continue to pollute and produce greenhouse gases for more than 50 years. These gases trap excess heat in our atmosphere, contributing to a rapidly warming climate.

According to the EPA, by 2030, the United States is expected to have as much as 1 million tons of solar panel waste. By 2050, the U.S. is projected to have the second-largest number of end-of-life panels in the world, with over 10 million total tons of panels. Currently, several Asian and European recycling companies are expressing interest in expanding operations in the U.S. Their goal is to extract valuable elements from these existing panels and find downstream uses. Disposing of outdated panels is a waste of valuable resources that could be repurposed to create new ones.

Given the industry's rapid growth, solar trash is becoming a looming issue. The lack of proper recycling infrastructure will lead to significant environmental problems if not addressed. For us to understand the solar waste crisis, we need to recognize that many current systems that are only 10 to 12 years old are already being decommissioned without fulfilling their full-life cycle. As they age, their efficiency decreases, necessitating replacement to maintain optimal energy production.

SOLUTIONS FOR MANAGING SOLAR WASTE:

Enhanced Recycling Technologies:
Developing advanced recycling methods can ensure that valuable materials are effectively reclaimed. For instance, innovations like high-voltage pulse fragmentation can separate raw materials more efficiently than traditional shredding methods.

Extended Producer Responsibility (EPR):
EPR policies require manufacturers to be responsible for the entire life cycle of their products, including disposal and recycling. Implementing such policies could encourage companies to design panels that are easier to recycle and reduce waste.

Legislative Framework:
Governments can play a crucial role by setting regulations for the proper disposal and mandatory recycling of solar panels. Such laws could ensure that all solar waste is handled responsibly.

Public and Private Sector Collaboration:
Collaboration between the government, private sectors, and research institutions can foster innovations in recycling technology and develop sustainable practices for end-of-life panel management.

Consumer Awareness:
Educating consumers about the importance of sustainable disposal and the potential environmental impacts of solar waste can drive public demand for more sustainable solar products and recycling services.

Given the predicted rise in solar panel waste, it is critical to design and execute effective recycling and waste management procedures. This includes increasing the effectiveness of recycling operations to recover valuable materials, reducing the environmental impact of hazardous substances, and encouraging a circular economy in which solar panel components are reused and repurposed. Furthermore, policies and regulations must be established to hold both producers and consumers accountable for solar panel end-of-life management. While solar energy provides a greener alternative to fossil fuels, managing solar waste is a new challenge that requires rapid and coordinated actions to avoid potential environmental consequences. Addressing this issue will be crucial for maintaining the benefits of solar technology while protecting the environment.

CHAPTER 3

DRIVING A CIRCULAR ECONOMY AND RECYCLING SOLAR MATERIALS

The concept of a circular economy has gained significant traction in recent years, especially within industries grappling with large-scale waste challenges. For solar energy, where the rapid deployment of photovoltaic systems is paralleled by an inevitable surge in end-of-life materials, adopting circular principles is not just an ideal—it is a necessity. Recycling solar materials effectively is central to this shift, ensuring that valuable components are reintroduced into the manufacturing cycle rather than being discarded. In this chapter, I share a personal story of innovation and adaptability in solar waste management—a challenge that required creative problem-solving and resulted in new best practices for transporting decommissioned solar panels.

OVERCOMING THE LOGISTICS OF SOLAR PANEL TRANSPORTATION:

Early in my efforts to manage solar waste, I encountered an issue that no guidebook or industry manual had prepared me for—how to safely and efficiently collect, strap, palletize, and transport thousands of broken solar panels. It was uncharted territory.

Even experienced installers and logistics companies were unfamiliar with the specific requirements of moving large volumes of damaged panels without risking further breakage or compromising safety.

I vividly recall the first time a 53-foot flatbed truck arrived on-site, ready to transport hundreds of broken solar panels. The driver stepped out, surveyed the towering stacks of broken glass and aluminum, and asked, "How are we going to fit all of this?" It was a valid question—at the time, there were no established methods or standards for handling solar panels in bulk.

This moment became a turning point. I realized that if we wanted to develop a sustainable approach to solar recycling, we needed to start with logistics. I began testing different ways to stack and secure panels, experimenting with pallet configurations, and calculating weight distributions to maximize truck capacity while minimizing damage during transit. After multiple trials, I developed a process that not only worked but became a template for others in the industry.

DEVELOPING BEST PRACTICES FOR TRANSPORTATION:

The methods we created for transporting solar panels have since become an integral part of our operations, ensuring safe, efficient, and sustainable logistics for solar waste management. Key elements of our best practices include:

1. **Strategic Pallet Stacking:**
   Panels are stacked in groups of 30 to 40, with careful attention to weight distribution. This prevents excessive pressure on the bottom panels and reduces the risk of breakage.

2. **Secure Strapping:**
   Heavy-duty, UV-resistant straps are used to tightly secure the panels to the pallets. This ensures that the stacks remain stable during transport, even over long distances.

3. **Protective Wrapping:**
   Some of the panels were cracked and very mangled. To prevent damage from road vibrations and debris, some of the pallets of panels are wrapped in durable protective plastic material. This additional layer of security reduces the risk of further damage to the panels.

4. **Load Optimization:**
   Truck space is meticulously planned to maximize the number of panels transported per trip without exceeding weight limits or compromising safety regulations. The driver will also use their ratchet straps to secure each section of the panels.

These practices have enabled us to transport solar panels across town and even across state borders efficiently and safely. What began as a logistical hurdle has now become a critical part of the broader effort to establish a circular economy for solar materials.

COLLABORATION WITH LOGISTICS COMPANIES:

Interestingly, this process has also involved educating logistics companies. Many truck drivers and transport coordinators had never handled solar panels before working with us. By sharing our methods and collaborating on-site, my company, Green Clean Solar helped create a new standard for solar panel transportation. Drivers who once hesitated now arrive with confidence, knowing exactly how to load and secure their cargo.

THE BROADER IMPACT:

The experience of solving this transportation challenge highlights an important aspect of driving a circular economy: innovation often starts with the practical problems that arise on the ground. By addressing these challenges head-on, we not only improve our own operations but also contribute to the development of industry-wide standards. These best practices for transporting solar panels are now being adopted by other recyclers and logistics providers, amplifying their impact.

As more industries adopt circular principles, the ability to adapt and innovate in response to new challenges will be essential. In the case of solar recycling, overcoming logistical barriers is just one piece of the puzzle—but it's a vital one. By ensuring that decommissioned panels can be transported safely and efficiently, we pave the way for their materials to be reclaimed, recycled, and reintegrated into the supply chain, creating a more sustainable future for solar energy.

The journey to a circular economy is filled with unexpected challenges, but each hurdle presents an opportunity for growth and innovation. My experience with solar panel transportation is a testament to the importance of hands-on problem solving and collaboration. By sharing what we've learned and continuing to refine our processes, we can drive meaningful change in solar waste management and set the stage for a truly circular renewable energy industry.

TACKLING THE CHALLENGES OF SOLAR BATTERY RECYCLING:

As the solar industry evolves, another pressing issue has emerged: managing the growing volume of solar batteries. Batteries used in solar systems present unique challenges for collection, transportation, and recycling due to their size, chemical composition, and potential hazards.

My company has taken a proactive approach to address these concerns, developing robust systems for safely managing thousands of end-of-life batteries.

Collection:
The process begins with meticulous planning to ensure all batteries are properly collected from decommissioned solar systems. We work closely with installers and site operators to identify and safely remove batteries, whether they are lithium ion, lead-acid, or another type. Each battery is inspected to assess its condition and determine the safest method of handling.

Packaging:
Proper packaging is critical when dealing with batteries, especially those that may be damaged or leaking. We use industry-approved containers designed to minimize risks during transport, such as fireproof and shock-resistant boxes. Batteries are carefully labeled according to their type and condition, ensuring compliance with transportation regulations.

Transportation:
Transporting batteries safely requires collaboration with logistics providers who understand the unique risks associated with these materials. Our team ensures that each shipment is securely packed, properly documented, and transported in compliance with all state and federal guidelines. This includes adhering to regulations set by the Department of Transportation (DOT) and the Environmental Protection Agency (EPA).

Recycling:
Once the batteries reach a reputable recycling facility, the materials are processed to recover valuable components like lithium, cobalt, and lead. These materials are then reintroduced into the supply chain, reducing the need for virgin resources and supporting the principles of a circular economy.

TRACKING AND REPORTING:

One of the most critical aspects of our battery recycling program is accountability. Each battery we collect is meticulously tracked from the point of collection to its final recycling destination. This tracking system allows us to generate detailed reports for our clients, demonstrating the environmental impact of their recycling efforts. These reports include information on the quantity and type of batteries recycled, the recovery rates of key materials, and the overall carbon footprint reduction achieved through the program.

SETTING INDUSTRY STANDARDS:

Our work in solar battery recycling is setting new standards for the industry. By prioritizing safety, sustainability, and transparency, we are helping to address one of the most complex challenges in solar waste management. As the demand for solar energy continues to grow, so too will the need for comprehensive solutions to manage the batteries that power these systems. By developing and sharing best practices, we aim to lead the way in creating a sustainable future for solar energy.

TACKLING THE CHALLENGES OF METAL RECYCLING:

There is a growing trend in the U.S. to restore steel production due to China's Zero-COVID policy and supply chain limitations. Similarly, the solar industry is witnessing an uptick in restoring initiatives, such as the NEXTracker & BCI partnership aimed at revitalizing an abandoned steel manufacturing plant in the historic Bethlehem Steel factory in Leetsdale. This new plant will manufacture solar tracker equipment for large-scale solar power plants, with production powered by renewable energy.

North America boasts one of the highest rates of steel recycling at 88%, contributing to the conservation of natural resources, emission reduction, and energy preservation. The European Parliament is exploring pathways toward zero steel or net-zero emissions in steel production. Additionally, the Climate Group and Responsible Steel have set ambitious targets of achieving 100% net-zero steel by 2050 and procuring 50% net-zero steel by 2030. The solar industry utilizes three types of steel, with hot-dip galvanized steel being the most widely used and highly recyclable. With metal reserves depleting, the solar industry is called upon to take a leading role in the metal recycling market. Recycled steel offers a significantly smaller carbon footprint compared to virgin steel and aids in the conservation of natural resources.

The Solar Industry heavily relies on aluminum, a metal that can be recycled numerous times without losing its properties, making it a popular material. This is why approximately 75% of all aluminum produced is still in use today, as it is constantly recycled.

The shortage of aluminum is not due to its rarity, as it is one of the most abundant elements on earth, but rather due to various reasons such as increased demand and caps on aluminum production and export by China. To alleviate the scarcity, the solar sector has to encourage more recycling and adopt recycling regulations for aluminum.

The demand for aluminum in the solar sector is projected to reach 486 Metric Ton by 2050, which could result in sending aluminum to landfills if recycling rates do not increase. Recycling aluminum has numerous benefits, including reduced carbon emissions and energy consumption. However, separating aluminum from other materials and finding the right re-processors can be challenging.

In order to ensure a steady supply of aluminum for the solar industry, it is crucial to improve waste management practices and embrace a zero-waste approach. One way to achieve this is by implementing a waste action plan before starting projects and making a commitment to recycle all aluminum which will help in diverting aluminum from landfills and reintroduce it into the recycling system. Ultimately, these efforts will benefit the industry in the long run by helping to maintain a steady supply and reduce costs.

The United States has numerous industrial scrap recycling facilities that can support these initiatives and promote domestic solar panel manufacturing.

The construction of large solar installation projects creates a significant amount of waste, including cardboard boxes used for shipping solar panels.

For example, a 74.5 MW facility used approximately 93,000+ cardboard boxes. Recycling cardboard offers several environmental benefits, such as reducing energy consumption and emissions. Furthermore, businesses and utilities can monitor their waste reduction efforts as part of their Environmental, Social, and Governance (ESG) reporting. Managing cardboard on-site can lead to cost savings and a reduction in the company's Scope 3 emissions.

Moreover, the large quantities of cardboard generated can be sold to re-processors for profit. Handling cardboard is relatively easy and can be done by the crew or contracted to a solar waste professional. The future trend is to reduce cardboard waste and use reusable systems, like P.V. Pallet's delivery system, to eliminate waste and improve sustainability.

CHAPTER 4

THE PROBLEM EXACERBATED:
THE ISLAND EFFECT:
RENEWABLE DEVELOPMENT OF WASTE MANAGEMENT

What is the significance of establishing sustainable energy and transforming solar recycling for island nations like **Hawaii**, the **Bahamas**, **Puerto Rico**, etc.?

Islands are known for their breathtaking landscapes and unique ecosystems. However, they face significant challenges when it comes to waste management and sustainability. Their limited land and proximity to vast oceans make waste management more complicated compared to larger mainland regions.

Islanders are currently focusing on responsibly managing waste as they transition to renewable energy sources in order to protect their natural environments and ensure a sustainable future. The move towards renewable energy is crucial for achieving sustainability. As islands reduce their reliance on fossil fuels, the importance of implementing solar and wind farms is growing. These renewable energy sources generate clean electricity, which decreases carbon emissions and reduces reliance on oil or nuclear power by preventing the release of greenhouse gases.

However, the rapid expansion of solar farms presents new challenges, as the handling of waste produced by these facilities needs to be closely monitored to avoid harm to the environment. Supporting renewable energy initiatives offers significant financial advantages to island communities. Solar and wind farms create employment opportunities in construction, maintenance, and related services and they promote energy independence by reducing dependence on foreign resources.

This change increases the stability and predictability of energy expenses, establishing a strong foundation for regional economies.

Furthermore, with careful planning, these efforts could reduce habitat disturbance and safeguard the diversity of indigenous ecosystems. Helping habitats to recover, creating wildlife pathways, and carefully choosing locations all assist in ensuring energy production and nature can peacefully coexist. Renewable energy projects on islands can enhance economic stability by cutting energy costs and reducing dependence on imported fuels.

Additionally, they support innovation and growth in other sectors of the economy. These projects also align with worldwide sustainability trends, helping island communities preserve their environment and remain competitive in the global market.

It is essential to involve the community and offer educational opportunities when working to gain support for renewable energy projects. Through involving local communities in the planning and development stages, a sense of ownership and responsibility is encouraged. Educational initiatives focused on renewable energy assist individuals in making well-informed decisions and adopting sustainable practices, ultimately resulting in a more environmentally conscious future.

Waste from packaging, predominantly plastic, poses a significant threat to the environment, particularly in island regions with restricted landfill space. Packaging, especially single-use plastics, has a significant impact on pollution, as plastic waste often finds its way into the oceans, harming marine life and disrupting ecosystems.

The manufacturing, transportation, and waste management of packaging materials lead to notable greenhouse gas emissions, adding to the environmental footprint. Islands face unique challenges in managing packaging waste. The limited land often results in landfills being filled beyond capacity, causing health risks and environmental damage. Additionally, it is crucial to maintain efficient waste disposal systems because of the island's proximity to the oceans, as improperly disposed packaging waste can directly impact coastal ecosystems.

Picture taken in **Maui** at a **Fruit Tree Landscape** project. Using **Solar Panel Boxes** to cover **Soil** to prevent **Weeds** and the use of **Pesticides**.

CHAPTER 5
OPPORTUNITIES - WORKFORCE DEVELOPMENT, JOB GROWTH IN SOLAR & RECYCLING

JOB GROWTH PROJECTIONS:

The outlook for job opportunities in the solar and recycling industries looks promising as projections show continuous growth. According to the Solar Energy Industries Association (SEIA), the U.S. solar industry alone could create as many as 40,000 new jobs each year for the next decade. Similarly, job growth in the recycling industry is expected to increase as more communities and businesses embrace sustainable waste management practices.

GROWING RECYCLING JOBS AND THEIR IMPACT ON THE SOLAR INDUSTRY

The rise of recycling initiatives across various industries has led to a surge in job creation, presenting a unique opportunity for the solar sector. As recycling operations expand to manage electronic waste, plastics, and metals, a skilled workforce is emerging with expertise in material recovery and waste management. This trend can have a transformative impact on the solar industry, where the need for efficient panel recycling is becoming increasingly urgent.

At Green Clean Solar, I am dedicated to revolutionizing the solar industry by merging these two sectors. By tapping into the talent pool from the recycling industry, we are building a workforce that understands the intricacies of material recovery and sustainable practices. Our approach not only addresses the growing challenge of solar waste but also provides meaningful employment opportunities in a rapidly evolving job market. This integration of expertise is paving the way for innovative solutions that will set new standards for sustainability and environmental stewardship in the solar industry. One of the things I'm most proud of is bringing skilled solar installers into the world of decommissioning jobs. Many of these talented professionals come from backgrounds focused on building new systems, and now they're using their valuable expertise to dismantle and repurpose old sites. Decommissioning isn't a task for just any labor crew—it requires skilled tradespeople and even electricians to ensure it's done safely and efficiently. Their previous experience in the solar industry makes them uniquely qualified to handle the complexities of these projects.

LEGISLATION AND POLICY SUPPORT

The employment dynamics within the solar and recycling industries are significantly shaped by government legislation. The implementation of supportive legislation for recycling and renewable energy on a global scale is driving the demand for a skilled workforce. For instance, the Inflation Reduction Act of 2022 in the United States has resulted in substantial tax incentives for solar energy installations and energy efficiency upgrades. According to the U.S. Department of Energy, these incentives are catalyzing the creation of employment opportunities across various domains, including system design and installation.

Similarly, the ambitious recycling targets and mandates stipulated in the European Union's Circular Economy Action Plan are fostering expansion within the recycling industry and prompting the need for a diverse array of new occupational roles (Source: European Commission). These regulatory measures have a cascading effect, leading to an increased demand for experts capable of executing and overseeing these initiatives. This impact may be more pronounced in island regions due to their limited resources and land availability. Supportive legislation can serve as a catalyst for small island nations and regions to adopt sustainable practices and generate local employment opportunities.

These regions often grapple with significant waste management challenges. For example, the Caribbean islands are swiftly enacting recycling programs and solar energy initiatives to address waste management issues and reduce their dependence on imported fossil fuels (Source: Caribbean Development Bank).

The metal scrap sector has long been a key player, employing thousands across the country to collect, process, and repurpose metals for new manufacturing. More recently, wind turbine recycling has emerged as a growing field, driven by the need to manage decommissioned turbine blades and other components sustainably. The push for renewable energy solutions and responsible resource management is fostering innovation and creating specialized roles in dismantling, sorting, and material recovery. This growth not only benefits the environment but also fuels economic development by providing skilled jobs and promoting technological advancements in recycling processes.

PHOTO OF **METAL BEAMS** LEFT OVER
FROM A **JOB SITE**

DIVERSE ROLES AND OPPORTUNITIES:

The rapid growth of the sustainability sector is creating a variety of roles designed to meet the specific needs of the solar and recycling industries. Each position plays a vital role in shaping a sustainable future, similar to individual threads coming together to create a vibrant tapestry. In the solar energy sector, we can identify:

Solar Panel Installers:

The individuals in question bear a striking resemblance to contemporary artisans, as they fastidiously oversee the installation and upkeep of solar panels. Their proficiency in electrical systems and safety regulations guarantees the smooth assimilation of sustainable energy into our daily lives.

System Designers:

With the precision of master planners, system designers craft solar energy systems tailored to specific energy needs. Their skills in engineering and technology enable them to design efficient and effective solutions.

Energy Analysts:

These analytical minds analyze data on energy production and consumption, optimize solar system performance, and assess financial implications. Their work forms the foundation of informed decision-making in the renewable energy sector. In the recycling sector, roles are equally vital.

Waste Management Professionals

These environmental stewards manage waste collection, sorting, and disposal to ensure compliance with regulations and responsible waste management.

Materials Recovery Specialists

Innovators in their own right, these specialists develop and implement technologies to recover valuable materials from waste streams, turning what was once discarded into something valuable.

Director of Sustainability

The role of Director of Sustainability has become a critical position for corporations across various industries as they increasingly recognize the importance of integrating environmental, social, and governance (ESG) considerations into their business strategies. This leadership role is responsible for developing and implementing sustainability initiatives that align with the company's goals, including reducing carbon emissions, improving resource efficiency, and fostering responsible supply chain practices.

With consumers and investors placing greater emphasis on corporate responsibility, the Director of Sustainability ensures that the organization meets evolving regulatory requirements while enhancing its brand reputation. By driving innovation and sustainable practices, these leaders help companies mitigate risks, identify new market opportunities, and create long-term value, solidifying sustainability as a core component of business strategy rather than an afterthought.

OPPORTUNITIES FOR ALL

As the sustainability sector expands, there is a focused effort to ensure that opportunities are accessible to all. Workforce development initiatives and inclusive hiring practices are being emphasized to diversify the workforce and provide training opportunities.

Programs like the U.S. Solar Training Network are dedicated to training and certifying individuals from underrepresented groups in the solar industry. Similarly, initiatives promoting diversity in recycling-related fields are helping to cultivate a more innovative and equitable workforce.

Island communities also stand to benefit from these inclusive approaches. By implementing regional training programs and creating pathways for diverse individuals to pursue careers in sustainability, islands can harness the full potential of their human resources. For example, the Caribbean Climate Innovation Center supports local job seekers and entrepreneurs in the recycling and sustainable energy sectors.

In the changing world of renewable energy and recycling, there is a coming together of different roles, strong projections for job growth, and opportunities for everyone. This creates a positive outlook for those interested in pursuing careers that contribute to a sustainable future, guaranteeing that the journey of progress and innovation will carry on.

THE ROLES OF COLLEGES AND UNIVERSITIES IN PREPARING THE WORKFORCE

Universities and colleges across the nation are increasingly incorporating sustainability-focused majors, minors, and certificate programs into their curricula. Recognizing the growing demand for a workforce educated in renewable energy, environmental stewardship, and sustainable business practices, higher education institutions are preparing students to address the challenges of a rapidly evolving green economy. These programs often combine science, policy, and technology courses to provide a multidisciplinary approach to solving environmental issues. Students are gaining hands-on experience through research projects, internships with renewable energy companies, and partnerships with sustainability-focused organizations. As I write this book, my oldest daughter, Alexis O'Leary, is a Junior and is studying at the University of Georgia with a major in Environmental Health Science while obtaining a Certificate of Sustainability. She is fortunate to have secured an internship with the Environmental Protection Agency (EPA), giving her invaluable real-world experience. This educational shift reflects a broader societal commitment to fostering a workforce equipped to build a more sustainable future.

CHAPTER 6

BEST PRACTICES FOR SOLAR PROFESSIONALS DEALING WITH SOLAR WASTE MANAGEMENT

Turning Solar Waste into Sustainable Opportunity: Essential Practices for Today's Solar Innovators

With the solar industry rapidly spreading through vast horizons, the responsibility to manage the resulting waste product becomes critical for all beings, especially solar professionals. Solar innovators are not only associated with driving the energy revolution movement, but they're also bound to ensure that the solar industry doesn't add to creating more environmental challenges for the world to suffer.

Proper waste management must be an integral part of every stage in the life cycle of a solar project. Neglecting this crucial aspect could diminish the environmental benefits of solar energy and contribute to greater pollution and resource depletion. Therefore, solar professionals are entrusted with the important responsibility of advancing clean energy while ensuring that their solar practices do not pose a threat to the environment. *The stakes are high!*

It's no surprise that solar projects generate significant environmental waste. From installation to decommissioning of the solar panels, it includes waste of all kinds for example, packaging material waste, damaged components, and end-of-life panels. Naturally, all such environmental solar waste requires proper waste management. Therefore, the challenge is clear: how can we continue to pursue creating solar energy while ensuring no added environmental problems? The answer is simple yet crucial: waste management. Integration of solar waste management practices into every stage of a solar project's life cycle is the key. From the design stage, which makes use of recyclable raw materials, to the disposal stage, which reduces landfill trash and recovers important resources. This is the approach that solar professionals should take to maintain a healthy community powered by the clean energy revolution.

PROFESSIONALS' DUAL RESPONSIBILITY

Solar professionals must balance clean energy generation with responsible waste management. Therefore, solar professionals need to focus on decarbonization while managing the waste generated throughout the life cycle of solar projects.

On one hand, solar energy is a cleaner and more sustainable form of energy. On the other hand, the methods of solar energy generation must not further harm the earth's atmosphere, which is already suffering. While the future of solar energy generation looks promising, it does come with the high cost of facing new environmental challenges. Solar professionals have a dual responsibility—not only to generate energy but also to ensure proper waste management as an integral part of the solar project life cycle. By embracing solar waste management practices, solar professionals can improve the long-term viability of their projects, reduce waste disposal costs, and contribute to a cleaner environment.

4 ESSENTIAL WASTE MANAGEMENT PRACTICES FOR SOLAR PROFESSIONALS

To fully leverage the environmental benefits of solar energy, the solar industry must adopt sustainable waste management practices. Exploring in-depth insights, strategies, and real-world examples can empower professionals to lead the development of a cleaner, greener future. As the solar industry expands, a comprehensive understanding of solar waste management is essential. Highlighted below are four key waste management practices that solar professionals can implement to minimize waste and promote sustainability.

DESIGN FOR SUSTAINABILITY

The implementation of sustainable waste management in solar product design is crucial for minimizing environmental impact. A key aspect in achieving this is the careful selection of materials that are not only durable but also recyclable and environmentally friendly. Solar companies should prioritize the reduction of waste by meticulously choosing raw materials for the design of solar panels and their components. Opting for materials with end-of-life recycling is essential to ensure a more sustainable approach, as it enables the recovery and reuse of materials when these products are no longer needed, thereby diverting them from landfills. Furthermore, the development of modular designs that facilitate easy upgradability and repairs plays a significant role in prolonging the lifespan of solar systems, thereby reducing the need for complete replacements. This approach not only contributes to environmental sustainability but also aligns with the principles of a circular economy.

RECYCLING AND REPURPOSING

Recycling and re purposing are the most crucial steps for managing waste in the solar industry. Since solar waste poses a great pollution threat, it's better to implement on-site recycling programs at solar plants during their installation and operation. This may help manage the solar waste effectively by ensuring that all the valuable materials, like metals or glass, are recovered at the end of the process and reused. Partnering with specialized recycling facilities would also be a prudent approach so that the solar waste can be processed properly, in compliance with standard environmental regulations.

WASTE MINIMIZATION DURING INSTALLATION

Minimizing waste production during the installation phase is another crucial aspect of sustainable solar practices. It can be the missing element to achieve greater solar project efficiency. By using materials more efficiently, planning installations carefully, and reducing excess waste, solar innovators can significantly reduce waste production and minimize its harmful effects on the environment. In addition, adopting cost-effective, returnable, and reusable packaging methods may also help decrease the amount of packaging waste generated during installations. This, in turn, contributes to reducing the carbon footprint and creating a more sustainable industry.

DECOMMISSIONING AND END-OF-LIFE MANAGEMENT

After a solar project's life cycle, it is imperative to emphasize the significance of comprehensive strategic planning. This phase plays a pivotal role in sustainable waste management due to its complexity. Developing a thorough decommissioning plan is essential to ensure the responsible dismantling of all solar installations after their operational lifespan. Prioritizing the recycling of materials and their safe disposal, as well as the recovery of valuable materials from decommissioned systems and the identification of safer methods for the disposal of hazardous substances, are critical steps in mitigating the environmental impact of solar projects.

The implementation of this approach guarantees that the advantages of solar energy persist even after its operational lifespan. These practices are imperative as they offer a systematic method for solar professionals to effectively handle solar installation waste in a more environmentally sustainable way, thereby fostering a beneficial influence on the environment.

LOOKING FORWARD TO A GREENER SOCIETY

The compelling necessity to implement optimal waste management practices within the solar industry is indisputable, given the imperative to mitigate environmental impact. Adaptation by solar professionals is crucial in reducing the carbon footprint associated with solar energy generation. The future trajectory of the solar sector is contingent upon its capacity for innovation, not only within energy generation techniques but also throughout the entirety of the solar project life cycle.

By adopting a proactive stance on solar waste management, professionals can be confident that their methodologies are in alignment with the fundamental sustainability principles upon which the industry was founded.

CHAPTER 7

A HIGHER CALLING –
SOLAR LEADERS ADOPTING
CIRCULAR ECONOMY PRACTICES
AHEAD OF MANDATES

The transition from *Megawatts* to *Mega Recycling* has been a growing concern in the solar industry. As I reflect on my own journey, I realize how essential it is for leaders like us to go beyond building energy solutions and look at the bigger picture.

It's about embracing a higher calling—one that involves adopting circular economy practices well before we are forced to do so by legislation.

When I first started in the solar industry, my primary focus was on helping companies tap into renewable energy. I was fueled by the promise of clean energy and the belief that we could make a difference by reducing carbon emissions. But as I installed system after system, it became clear that our responsibility didn't stop there. The solar panels, while a marvel of sustainable technology, have a finite lifespan. And what happens after their usefulness has expired?

The truth is, without a proper end-of-life plan, the solar industry risks becoming another source of significant waste. And that's why I founded Green Clean Solar—not just to provide waste management services, but to be part of a solution that's forward-thinking, responsible, and grounded in sustainability. We have the technology, the knowledge, and the will to close the loop on solar waste.

The circular economy, where resources are reused, recycled, and reintroduced into the manufacturing process, ensures that we're not only reducing waste but also preserving vital materials for future generations. By adopting these practices, we're taking ownership of the entire lifecycle of our products, not just the part that's profitable today. And that shift in mindset is what will set us apart as true leaders in renewable energy.

What drives me the most is thinking about my children—and the next generation. They will inherit the world we're shaping now. Every time we fail to recycle a panel, a battery, or a component, we're leaving behind a problem they'll have to fix. But by building a circular economy today,

We're giving them something much more valuable: a sustainable future.

For us, solar isn't just about reducing dependence on fossil fuels anymore—it's about leaving no trace, ensuring that every panel we produce, every component we use, is accounted for and given new life. The responsibility we have as solar leaders extends beyond innovation. We must pioneer a new standard in the way we handle the materials we rely on, setting the example for industries to follow. And the time to act is now, before we're required by mandates, because by then, it might be too late to lead the change.

It's not always the easiest road. Transitioning to circular economy practices can be complex and costly upfront. But the long-term benefits, both environmental and economic, are immeasurable. We're keeping valuable materials—like glass, silicon, and rare metals—in circulation, reducing the need for mining and raw material extraction, which in turn lowers carbon emissions. We're cutting down on landfill waste and the environmental degradation that comes with it. And, most importantly, we're creating a renewable energy future that's truly sustainable.

By embracing a circular economy today, we ensure that solar energy is not just a solution for the present, but a legacy for the future. Solar leaders must be bold enough to take on this challenge, because the future of our planet—and our children's world—depends on it. Let this be our higher calling, a responsibility we carry as pioneers of clean energy, stewards of the environment, and guardians of the future.

CHAPTER 8
FUTURE TRENDS OF
RECYCLING RENEWABLES

As I sit down to write the final chapter of Megawatts to Mega Recycling, I can't help but feel hopeful. It's been a journey of reflection, innovation, and action. We've uncovered the challenges, discussed the pitfalls, and mapped out the obstacles, but now it's time to turn our gaze toward the future—a future that, in my heart, I believe will be defined by transformation and progress.

The renewable energy industry has always been forward-looking. We're in this because we believe in the power of tomorrow. And when I look at the trends emerging in recycling solar, wind, and other renewable, I see something remarkable: A movement toward sustainability that's unstoppable.

In the past, we approached recycling as an afterthought. The focus was on generating as much clean energy as possible, and we hoped the rest would sort itself out. But now, we're entering a new era, One where waste management, material recovery, and circular economy principles are embedded in the DNA of every renewable project. This isn't just about compliance or cost; it's about embracing the full responsibility of what it means to create sustainable energy systems.

I've seen the shift firsthand in my own company, Green Clean Solar. What was once a niche service—recycling solar and wind waste—has become a critical component of our industry's road map. More and more, we're not just decommissioning sites;

we're ensuring that nothing goes to waste. The glass, the silicon, the metals—they're all finding new life, being reused and reintroduced into the supply chain. That's the beauty of the future: it's regenerative.

One of the most exciting trends I've noticed is the advancement in recycling technology. Researchers and engineers are developing methods to recover more materials from solar panels and wind turbines than ever before. We're talking about processes that extract nearly 100% of valuable components, which were once considered impossible to recover. The efficiency of recycling is improving, and this is changing the economic equation—turning waste into a valuable resource.

In the coming years, we'll see solar panels designed with recyclability in mind. Manufacturers are beginning to embrace the concept of designing for disassembly, creating panels that are easier to take apart, with components that are simpler to separate and reuse. This will significantly reduce the environmental impact of manufacturing and disposal.

It's not just solar. Wind turbines, batteries, and energy storage systems are all undergoing similar transformations. As demand for clean energy continues to grow, so does the need for a system that recycles every component. Companies are realizing that their success is tied not only to how much renewable energy they can produce, but also to how efficiently they can recover and repurpose their materials.

But what excites me most about the future is the shift in mindset. More than ever, there's a recognition that this isn't just a challenge for governments or corporations—it's something we all own. Communities, entrepreneurs, and leaders at every level are stepping up, understanding that the future of renewable energy is as much about what we do with our waste as it is about the energy we produce.

The younger generation, in particular, is demanding accountability. They want clean energy, yes—but they also want to ensure that the systems creating it are as sustainable as possible. I see the passion in my own children. They're growing up in a world where sustainability isn't just a talking point—it's a way of life. And they expect that from us, their parents, their leaders, their industries. They don't see the world in silos of energy production and waste. They see the whole picture, and they're challenging us to be better. To me, that's the most positive trend of all.

As we move forward, I'm filled with optimism. The road ahead isn't without challenges, but it's clear that we're heading in the right direction. We're learning, evolving, and innovating. We're building a future where renewable energy is not only abundant but endlessly renewable itself—where every panel, every blade, every battery can be recycled, repurposed, and reused, over and over again.

That's the legacy I want to leave behind. A future where the term "mega waste" doesn't exist—because we've found ways to turn it all into something "Mega Recycling". A future where renewable energy truly lives up to its name. Where our children and their children can look back and see that we didn't just solve the energy crisis, but that we did so with care, foresight, and respect for the planet they call home.

THIS IS THE **FUTURE** OF **RECYCLING RENEWABLE S.** AND I COULDN'T BE MORE **EXCITED** TO BE **PART OF IT.**

CONTINUING THE JOURNEY:

While **Megawatts to Mega Recycling** has explored the foundational shifts towards a more circular economy, the challenges of waste management continue to demand innovative solutions. My next work, **Megawatts to Mega Landfill Diversion**, builds directly on these principles, offering an in-depth look at transformation strategies for diverting vast quantities of waste from landfills and realizing the full promise of a resource-rich future.

ABOUT THE AUTHOR

Emilie Oxel O'Leary is first and foremost an entrepreneur with a pioneering force in the solar and renewable energy industry, dedicated to revolutionizing waste management and sustainability.

As the **CEO** and **Owner** of **Green Clean Solar**, she leads a team committed to tackling the growing issue of solar and wind waste, focusing on landfill diversion, decommissioning projects, and sustainable recycling solutions. With decades of experience in the industry, Emilie has built a reputation as a trailblazer, advocating for responsible end-of-life practices in renewable energy while fostering a diverse workforce. Before founding Green Clean Solar, Emilie successfully built and sold a solar mechanical installation company. Her company became a trusted partner for large-scale solar projects across the nation, earning recognition as master installers for Fortune 500 companies, universities, and military bases. Her early career also includes a high-profile solar carport installation at the Mercedes-Benz Stadium in Atlanta—a project that shaped her commitment to sustainability and innovation.

Emilie's impact isn't confined to her entrepreneurial accomplishments; she's a dynamic force reshaping the clean energy landscape. Through thought leadership, hands-on collaboration, and her unwavering dedication to mentoring future innovators, she champions a smarter, more sustainable future.

She frequently speaks at conferences and mentors the next generation of clean energy leaders, including students at the University of Georgia. Furthermore, she works closely with EPCs (Engineering, Procurement, and Construction companies), owners, and manufacturers to advance responsible recycling practices. Her mission is to ensure that the renewable energy sector not only produces clean power but also upholds its promise of environmental stewardship. *Megawatts to Mega Recycling* is her call to action— a deep dive into the often-overlooked reality of solar waste and the urgent need for sustainable solutions in the industry.

---

[i] Emilie Oxel O'Leary, Personal Project

A LETTER FROM EMILIE

Dear Reader,

Thank you so much for taking the time to read *Megawatts to Mega Recycling*. It truly means the world to me that you joined me on this journey through the often-overlooked side of the solar industry what happens when renewable energy equipment reaches the end of its life. My hope is that you walked away with new insight, perhaps even a deeper understanding of the environmental challenges and the massive opportunities we have to rethink, repurpose, and recycle in a cleaner, smarter way.

www.Emilieoleary.com

If this book sparked your curiosity or inspired you to take action whether you're in the energy industry, a sustainability advocate, or simply someone who cares about the future of our planet, I'd love to stay connected. Please follow me on LinkedIn at Emilie Oxel O'Leary, where I regularly share updates, industry news, and stories of progress in circularity and clean energy solutions.

And if you want to reach out directly, I welcome your questions, ideas, and partnerships. You can email me anytime at Info@emilieoleary.com

With gratitude,
Emilie Oxel O'Leary

www.ingramcontent.com/pod-product-compliance
Lightning Source LLC
Chambersburg PA
CBHW070807260626
47161CB00006B/2188